W9-DFO-090

NUCLEAR Disasters

Rob Alcraft

Heinemann Library
Chicago, Illinois

Published by Heinemann Library,
an imprint of Reed Educational & Professional Publishing,
100 N. LaSalle, Suite 1010
Chicago, IL 60602

Customer Service 888-454-2279

Designed by Celia Floyd
Illustrations by David Cuzik (Pennant Illustration) and Jeff Edwards
Originated by Dot Gradations
Printed by Wing King Tong, in Hong Kong

04 03 02 01 00
10 9 8 7 6 5 4 3 2 1

Library of Congress Cataloging-in-Publication Data

Alcraft, Rob, 1966-
Nuclear Disasters / Rob Alcraft.
p. cm. – (World's worst)
Includes bibliographical references and index.
Summary: Examines the events leading to disasters at nuclear power plants in Great Britain, the United States, and the Soviet Union, the consequences of these accidents, and how they might have been averted.
ISBN 1-57572-989-X (library binding)
1. Nuclear power plants—Accidents Juvenile literature.
[1. Nuclear power plants—Accidents.] I. Title. II. Series.
TK9152.A547 1999
363.17'99-dc21 99-27879
CIP

Acknowledgments
The Publishers would like to thank the following for permission to reproduce photographs:
Science Photo Libarary/U.S. Dept. of Energy, p. 4; Telegraph Colour Library/Bildagentur Scenics, p. 6; Corbis/Hulton-Deutsch, p. 9; Rex Features/Jorgensen, p. 12; Halstead, p. 15; Media Press International, p. 19; Topham Picturepoint/Unep, p. 21; Shone/SIPA, p. 24; Gamma/Hergott/Contrast, p. 25; Greenpeace/Core, p. 26; Greenpeace/Deiman, p. 27.

Cover photograph reproduced with permission of David Harden/Impact.

Every effort has been made to contact copyright holders of any material reproduced in this book. Any omissions will be rectified in subsequent printings if notice is given to the Publisher.

Some words are shown in bold, **like this**. You can find out what they mean by looking in the glossary.

Contents

Nuclear Power

Electricity is such a big part of our lives. We flip a switch and expect instant power. But did you know that when you use your television or toaster you could be using electricity from a **nuclear power station**?

One-fifth of the electricity we use is made in nuclear power stations. There are over 400 across the world. The very first was a small **prototype**, built in 1951.

Just 2 pounds (1 kilogram) of uranium can make as much heat as 6,600,000 pounds (3 million kilograms) of coal. This lump of uranium weighs about 10 pounds (4.5 kilograms).

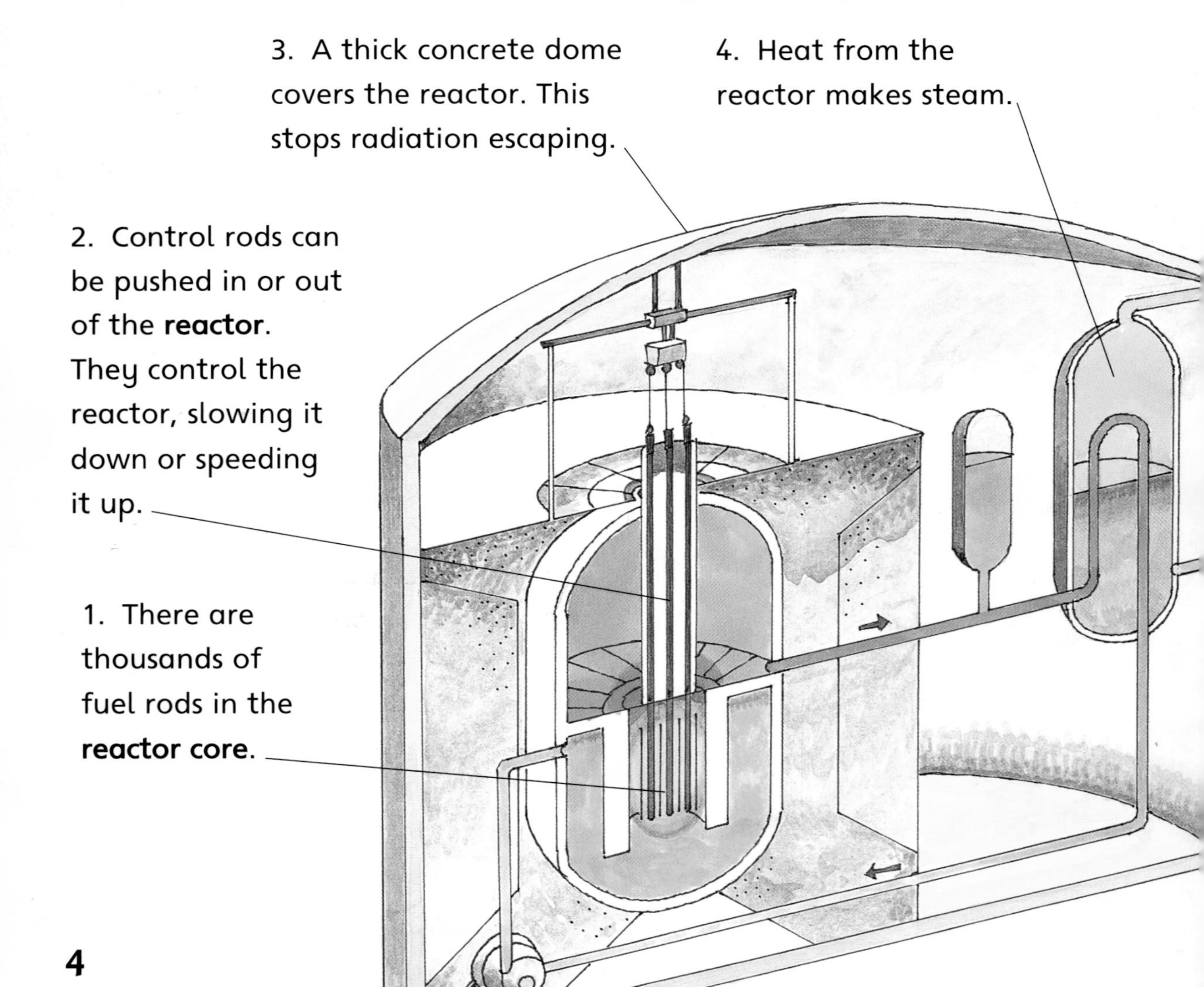

Making electricity

Nuclear power stations work in much the same way as ordinary power stations. They heat water to make steam. The steam turns **turbines** and **generators**. The generators make electricity. But instead of burning gas, oil, or coal to make heat and steam, nuclear power stations use **uranium** fuel rods. These give off energy and heat, or **radiation**.

Nuclear power stations can produce electricity very cheaply. Yet they are very expensive to build. Each new nuclear power station costs billions of dollars. Because of these costs, and the fear of accidents, very few new nuclear power stations are being built today.

Energy from atoms

Atoms are some of the smallest things we know about. You can imagine atoms as tiny building blocks. They group together to make you and everything around you.

Atoms are held together by energy. If an atom is split or knocked, it lets go of the energy. In nuclear reactors, atoms are split on purpose so that they will let go of their energy and make heat. Splitting atoms to make energy is called **fission**. Fission is what happens inside a nuclear reactor.

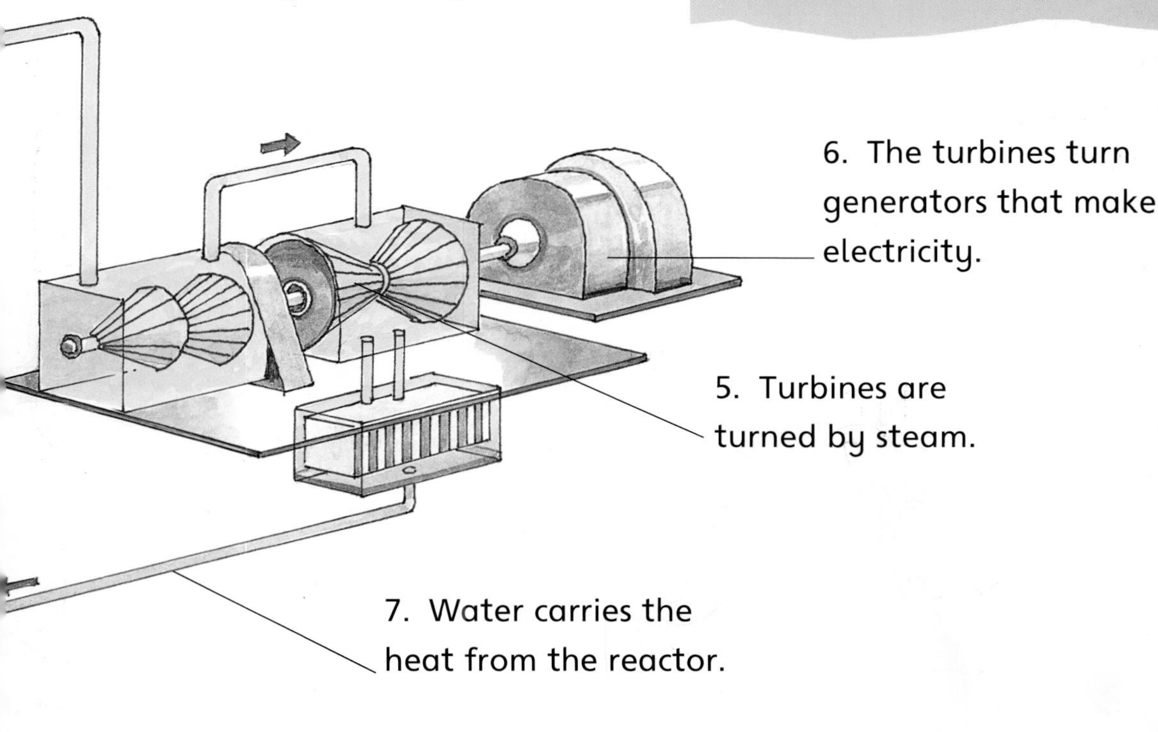

When Things Go Wrong

In this book, we look at three of the world's worst nuclear accidents, where **nuclear power stations** have burned, exploded, or leaked **radioactive** waste into the land around them. We look at what went wrong and what we can learn to make the world safer.

Built-in danger

The **uranium** fuel in today's **reactors** will still be dangerous in 150,000 years. If it should escape or leak, it will kill or poison every living thing over a huge area. Even small amounts of **radiation** are dangerous and can increase the risk of **cancer**. High doses of radiation can kill.

Because nuclear power is so dangerous, great care is taken when nuclear reactors are designed and built, so that accidents won't happen. Strict rules are set by international authorities, and many countries have an organization to oversee nuclear power, such as the Nuclear Regulatory Commission (NRC) in the United States. But nuclear power stations are still run by people—and people make mistakes. When they do, disaster can follow.

Other nuclear accidents are less dramatic, but still serious. Radioactive material can escape from the reactor. It can leak through **valves** or broken pipes and get into the steam, water, or air that is used to carry heat from the reactor core to the **generators**. From here it can get into the **environment** where it can poison food, soil, and people.

Nuclear accidents

Nuclear reactors cannot explode like nuclear bombs. But if they are not cooled properly, they can overheat and burn. Called a **meltdown,** this is the worst sort of nuclear accident. In a meltdown, the **reactor core** turns into red hot liquid. It can cause explosions in the machinery around it. It can burn through concrete shields. Large amounts of radioactive material can escape.

The destructive power of a nuclear bomb comes from splitting atoms. This same power is used in nuclear power stations to make heat.

Nuclear waste

Nuclear waste includes used **fuel rods** and leftovers from making nuclear weapons. This waste stays dangerous for over 10,000 years. One kind of nuclear waste, a metal called **plutonium**, is the most dangerous substance on Earth. One millionth of a gram is enough to cause cancer. No one has yet decided what to do with the world's nuclear waste. It is stored—usually at nuclear power stations—while governments work out what to do with it.

Red Alert!

Fire at Windscale Nuclear Power Station

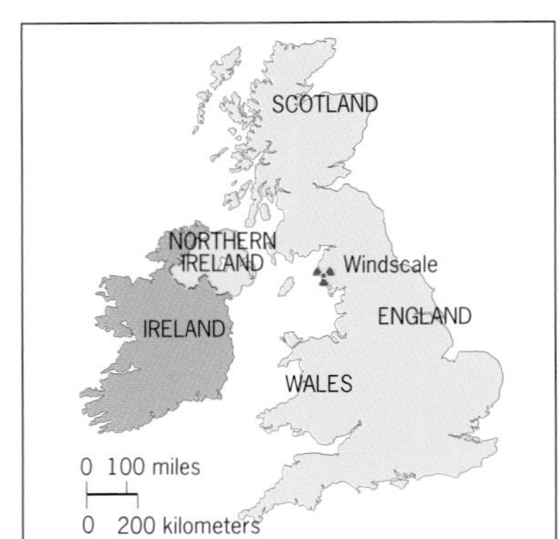

On October 10, 1957, a **reactor** at Windscale **nuclear power station** in England burst into flames. **Radioactive** dust drifted in clouds from the power station. With the **reactor core** burning red hot, Windscale was on the verge of **meltdown**.

The reactor catches fire

The problems started during normal maintenance on October 8. The reactor at Windscale was heated up to get rid of a buildup of energy inside. Operators tried to keep the reactor at 625°F (330°C), the ideal temperature for the job. Yet the controls seemed slow. The reactor wasn't responding as they expected. By the afternoon of October 9, the temperature inside the reactor had risen to over 752°F (400°C). But the operators did not know this. They could only see the temperature in part of the reactor, not all of it.

It wasn't until 2:30 P.M. on October 10 that the operators realized that something, somewhere, was wrong. Windscale was burning.

Ron Gausden, in charge of the heating-up process, gave orders for his men to push the **fuel rods** from the burning reactor core to keep the fire from spreading. But in the heat of the fire, the fuel rods had melted and bent. Some seemed stuck.

Eight brave men, after having dressed in protective suits and breathing masks, struggled to remove the plugs over each fuel channel.

Using crowbars and scaffolding poles, the men pushed the 120 burning fuel rods down their channels and away from the fire. As the men moved from rod to rod, the crowbars came out dripping with **molten** metal.

> *Nobody showed any signs of fear. You couldn't have seen a better display from the process workers [who moved the burning fuel rods]. They were heroes that night.*
>
> Chief Fire Officer, Windscale, 1957

Although there was no explosion at Windscale, enough **radiation** escaped to kill several people and make many more ill.

Fire and Luck

At 7:00 P.M., Tom Tuohy, Windscale's deputy manager, climbed up onto the **reactor's** roof. Peering through a glass-covered hole called an inspection port, he could see the glow of fire. An hour later there were yellow flames. At 11:30 P.M., Tuohy could see blue flames—a sign that the fire was getting hotter and was spreading. Managers at the power station had never dealt with a runaway fire in the reactor before.

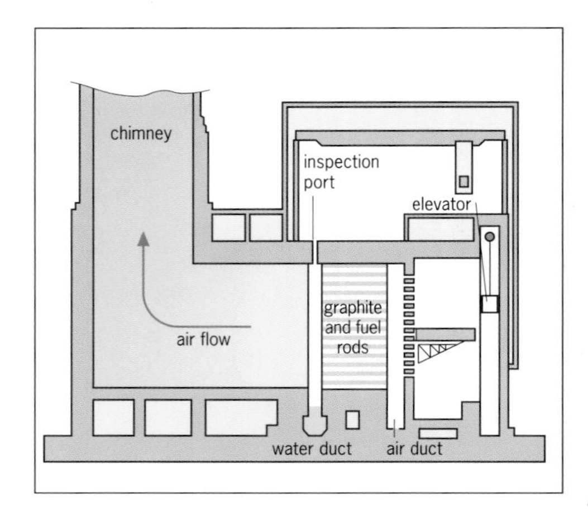

1. On October 8, the events leading to disaster begin. Operators try to release heat and energy from **graphite rods** that control the reactor. They don't realize that there is a large risk of the reactor overheating.

2. By October 10, the reactor is overheating. It begins to burn. Eight men struggle to push **fuel rods** along their channels away from the fire. The fire stops spreading, but it does not go out.

3. At around 4 A.M., carbon dioxide (CO_2) gas is pumped into the reactor. Managers hope that this will smother the fire—but it burns on.

6. On October 11, the wind over Windscale changes. **Radioactive** dust is blown inland. Emergency tests on milk from cows in the area show dangerous levels of radioactive material. But it is four days before proper warnings are given to people living around the power station. All milk is banned from sale over a 12,500-acre (5000-hectare) area. Farmers pour it into ditches.

5. At 9:00 A.M., water hoses are turned on. It is only a trickle of water, but it could be enough to cause an explosion. Managers watch tensely. Gradually, as the hoses are turned up, it is clear that Windscale will be safe. The hoses run for 30 hours.

4. By the early hours of October 11, managers at the plant decide to use water on the fire. This method is dangerous and untested—it could cause an explosion. But there is nothing else left to try. They lower hoses into position above the reactor.

Windscale Secrets

The disaster at Windscale was followed by investigations and reports by experts. But ordinary people were told almost nothing. It was 26 years before the facts—about what happened and what could have happened—were made public.

The government estimates that 33 people died because of the Windscale disaster. This is the number of extra deaths in the area thought to have been caused by the **radiation**. There were also over 200 extra cases of **cancer** caused by Windscale.

The real facts about how many people suffered because of Windscale will never be known. It is hard to tell whether people are ill because of radiation or for some other reason. The effects may take years to show. Children can become ill because their parents were harmed by Windscale—but proving this is very difficult.

In the end, what really saved Windscale from **meltdown** was the people who worked there, including those who struggled to move the **fuel rods** out of the way of the fire.

The Report of the Penny Inquiry, which investigated the disaster, said this about the Windscale workers:

The steps taken to deal with the accident, once it had been discovered, were prompt and efficient and displayed considerable devotion to duty on the part of all concerned.

After the Windscale disaster, scientists attempted to measure how much radiation had escaped by studying soil samples.

A disaster waiting to happen

Windscale was one of the first **nuclear power stations** ever built. It was designed and built in just four years. The British government wanted it quickly because it would produce a nuclear waste called **plutonium**. Plutonium is used to make nuclear bombs. The British wanted to keep up with the U.S. and the **USSR**, countries that had built and tested nuclear bombs.

The **reactors** at Windscale were unstable and unpredictable. They used **graphite rods**—like very thick pencil leads—to control the power of the reactor. But over time, these **control rods** became swollen and bent. Like batteries, they stored heat and energy. Every so often this energy had to be released by heating the reactor. But it was dangerous. Even if the energy was released intentionally by the reactor's operators, the heat could run out of control. On October 10, 1957, that is exactly what happened.

Burn Up!

Disaster at Three Mile Island Nuclear Power Station

CANADA
USA
PENNSYLVANIA
Three Mile Island
0 600 miles
0 1000 kilometers

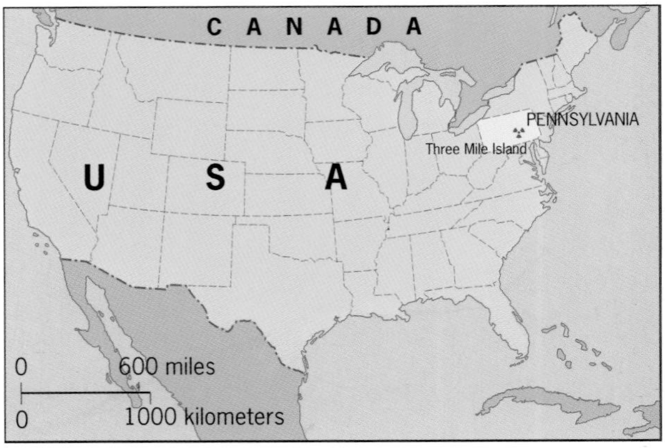

In the early hours of March 28, 1979, a reactor at Three Mile Island nuclear power station in Pennsylvania began to overheat. It seemed that meltdown was only minutes away. There was panic across the island. No one died, but it is still not really known whether there has been a significant effect on people living in the area.

Flashpoint!

It was 4:00 A.M. Suddenly the sirens began wailing in the power station control room. A pump feeding the reactor's **cooling system** had failed.

The operators kept calm. They had been trained for emergencies. Pumps had failed before and the reactor had stayed safe. But a few minutes later, red warning lights began to flash. The entire cooling system in the reactor had failed.

In just 15 seconds, the temperature in the reactor had rocketed by 540°F (300°C). Emergency pumps should have been pumping cooling water around the reactor. Nothing happened. Now the operators weren't so calm. In the confusion, another emergency pump was accidentally shut down. The temperature in the reactor reached 4900°F (2700°C). The reactor's protective walls fell apart. **Radioactive** steam spewed from a waste tank into the **environment**. All of this happened in eight minutes.

Thousands of people fled from Three Mile Island when news of the disaster came out. They were afraid of the **radiation** that might escape from the power station.

Mystery bubble

The fire in the reactor caused a mystery bubble to form in the reactor building. The bubble contained hydrogen, an explosive gas. It could **ignite** and blow the roof off the reactor. But as experts tried to decide why it was there and how to deal with it, the bubble began to shrink on its own. Luck had prevented an even greater disaster.

Emergency

It took just eight minutes for the **reactor** at Three Mile Island to reach **meltdown** temperature of 6500°F (3600°C). A series of instruments, computers, and **valves** failed to work. People made mistakes. "It seemed to go on and on, surprise after surprise," said **radiation** protection supervisor Thomas Mulleavy.

Two protective concrete shells covered the reactor. These were meant to hold in any explosions or leaks of **radioactive** material. The reactor burned through one shell, spilling large amounts of radioactive fuel and gases into the outer shell. Some of these escaped into the **environment**. If the reactor had burned through the outer shell, many lives would have been lost.

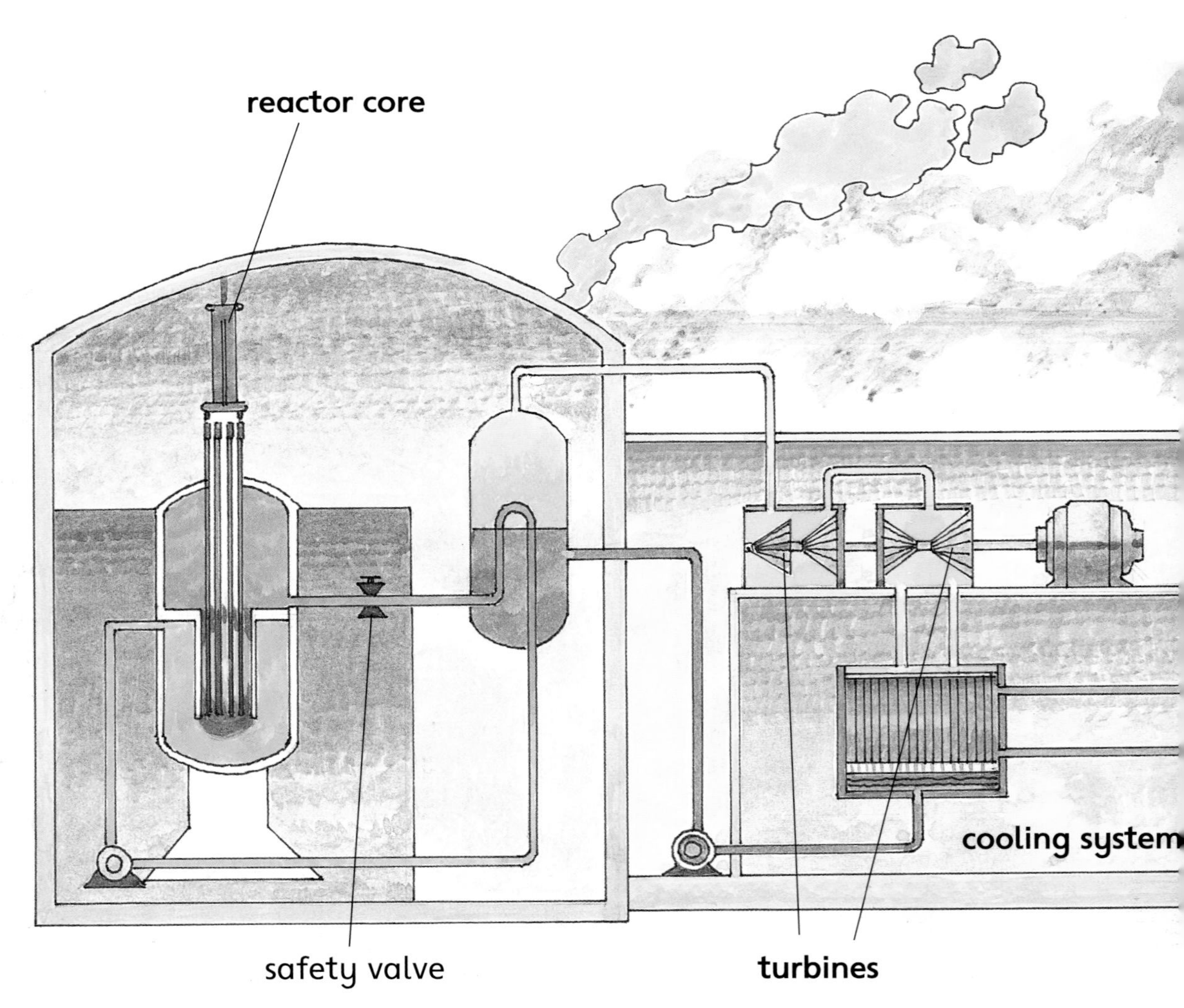

1. At 4:00 A.M. on March 28, 1979, a water pump fails. The reactor begins to overheat. The emergency cooling system also fails. In minutes the reactor begins meltdown.

2. Operators realize that vital valves aren't working. They trigger these by hand. Water floods through into the reactor. The reactor is saved from meltdown. It is too late to keep a radioactive cloud from drifting towards the nearby city of Harrisburg.

3. Water used to cool the reactor leaks into the reactor building. The water is highly radioactive. A 400,000 gallon (almost 2 million liters) waste tank of less radioactive water has to be dumped into the Susquehanna River to make room for it.

4. The emergency is made public at 11:00 A.M. on March 28. The telephone system is jammed. Troops move in and 140,000 people flee the island. President Carter declares a **national emergency**. Children and pregnant women living within 8 miles (13 kilometers) of the power station are most at risk from the radioactivity. They are told to leave.

What Went Wrong?

It was two weeks before the Three Mile Island **nuclear power station** was declared safe. Then the questions began. Why had it happened? Had the disaster been an accident? Was someone or something to blame?

The $1 billion mistake

The disaster at Three Mile Island changed everything for nuclear power. People found out that nuclear power wasn't as cheap, clean, or safe as experts had said it was. Many no longer trusted this new technology.

New safety rules were established, including fire equipment, new emergency plans, extra training for operators, and new inspectors for each nuclear power station. Many nuclear power stations had to shut down. They couldn't afford the cost of the new rules. Orders for new nuclear power stations were canceled.

There was a record of near accidents at the Three Mile Island plant. Twice before, important **valves** in the **cooling system** had failed to work. These were the same valves that were involved in the disaster. It became clear that although the power company that ran Three Mile Island power station had known there was a risk, they had failed to act.

Experts also found that the instruments the operators relied on to tell them what was happening in the **reactor** were very slow. Sometimes they gave readings that were one and a half hours out of date. Operators would not know how bad an emergency was until it was too late.

Three Mile Island also had another problem. The nuclear power station was expensive to build. In order to recover some of this money, Three Mile Island ran almost constantly. Even safety checks were conducted as the plant ran. Repairs were often left until official shutdowns so that no money-making electricity would be lost. Safety at Three Mile Island was ignored too often. In the end, this led to disaster.

The China Syndrome

Not long before the disaster at Three Mile Island, movie-goers were watching a new film called *The China Syndrome*. It was about a nuclear accident. When experts from the American nuclear industry saw the film, they said the accident in the film couldn't happen. At Three Mile Island, it did.

In the film *The China Syndrome,* two TV news reporters (played by Jack Lemmon and Jane Fonda) chase a story about a scare at a nuclear power station.

Meltdown!

The Chernobyl Nuclear Power Station Disaster

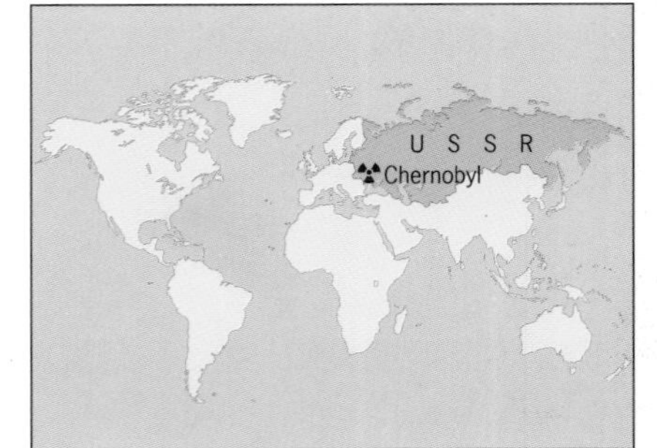

On April 26, 1986, a **reactor** at the Chernobyl **nuclear power station** exploded. A deadly **radioactive** cloud spewed from the burning reactor. Hundreds of people were killed. Two hundred thousand people were affected. Even today, people are dying from **radiation** unleashed during the disaster.

Steps to disaster

On April 25, 1986, scientists at the Chernobyl nuclear plant in the **USSR** began a safety test on reactor number 4. During the test, the reactor's emergency **cooling system** was switched off.

In the control room, the operator began the test. He slowed the reactor and reduced power. The emergency cooling system was disconnected. At 2:00 P.M., electricity workers in nearby Kiev requested that the test be delayed. They needed electricity. The Chernobyl staff brought the reactor back up to normal power. But the emergency cooling system was not turned back on.

This was the first of a string of serious errors. At 1:23 A.M. the next morning, the situation became critical. Reactor power surged. Extreme heat began to break up the reactor from the inside. It was reaching **meltdown**, and there was nothing anyone could do to stop it. Machinery, pipes, and tanks around the reactor exploded.

Showers of burning concrete and deadly lumps of reactor tore through the roof. Dozens of fires burned. Clouds of deadly **radioactive** gas and smoke drifted up into the sky.

This wreckage is what remains of the reactor after the explosion at Chernobyl. Standing near it, even for a few minutes, meant absorbing life-threatening doses of radiation.

We heard heavy explosions! You can't imagine what's happening here with all the deaths and fire. I'm here 32 kilometers [20 miles] from it, and in fact I don't know what to do. I don't know if our leaders know what to do, because this is a real disaster. Please tell the world to help us.
(Transmission received by Annis Kofman, Dutch radio operator, on April 29.)

Heroes of the fire

Many of the firefighters, soldiers, pilots, and doctors on duty at Chernobyl knew the dangers of radiation. They knew that without protective suits or breathing masks they would become ill or die. There was no protective clothing available, but they carried on anyway. In under an hour many were already too ill to continue their efforts.

Nuclear Inferno

Politicians and officials had said that a disaster would never happen at the Chernobyl **nuclear power station**. When it did, no one was ready. Fire crews and soldiers had no protective suits or breathing masks to protect them from **radiation**. For ten days they battled with the immense heat of the burning **reactor**.

1. The disaster at Chernobyl begins on April 25, 1986, with a safety test on the reactor. Operators make serious mistakes, and cooling pumps are operating beyond their capability. At 1:19 A.M. on April 26, the test should have been stopped. But operators override the automatic shutdown and carry on. There are only seven or eight **control rods** left in the reactor. The safe minimum is 30.

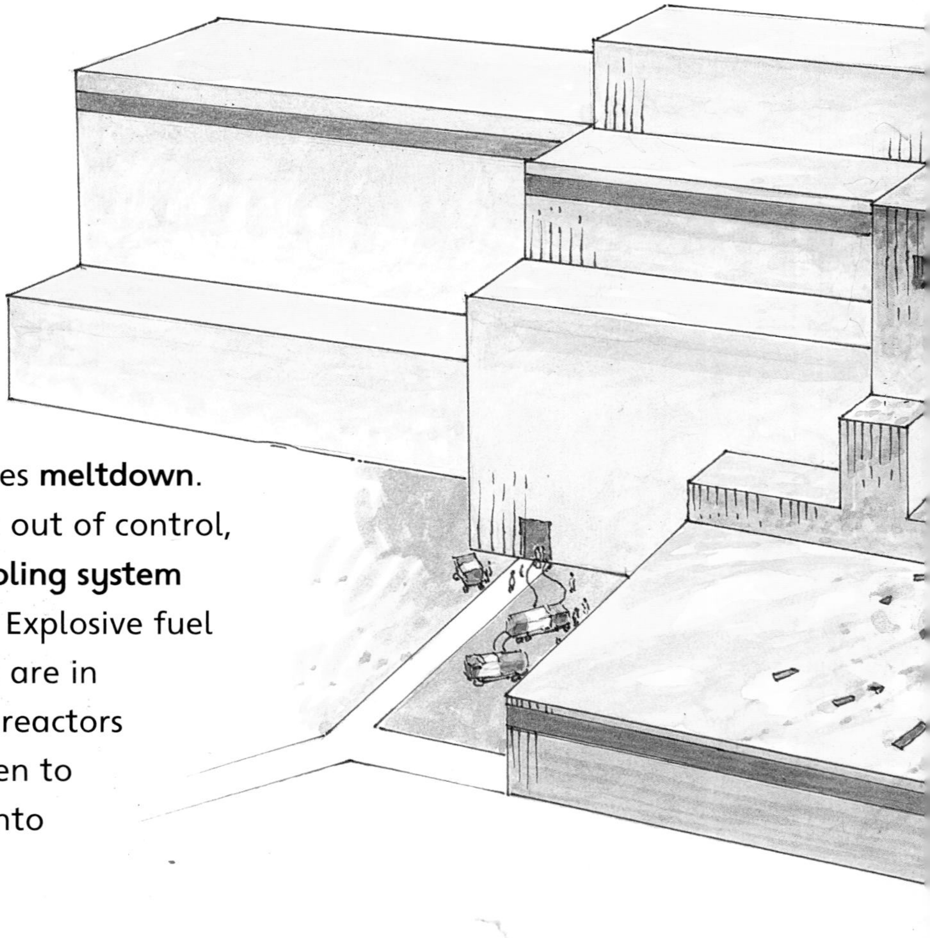

2. The reactor reaches **meltdown**. Temperatures rocket out of control, and the reactor's **cooling system** explodes. Fires start. Explosive fuel and hydrogen stores are in danger. Three other reactors at Chernobyl threaten to burn. Fire crews go into action at 1:30 A.M.

5. A poisonous cloud of radiation floats over the Ukraine and on into Europe. The wind and rain deposit radiation as far as Wales. In Greece, people are afraid of radiation poisoning. In Germany, milk sales are banned because cows are eating radioactive grass. In Italy, tons of fresh vegetables are dumped because of fears that they are **contaminated** with radioactivity.

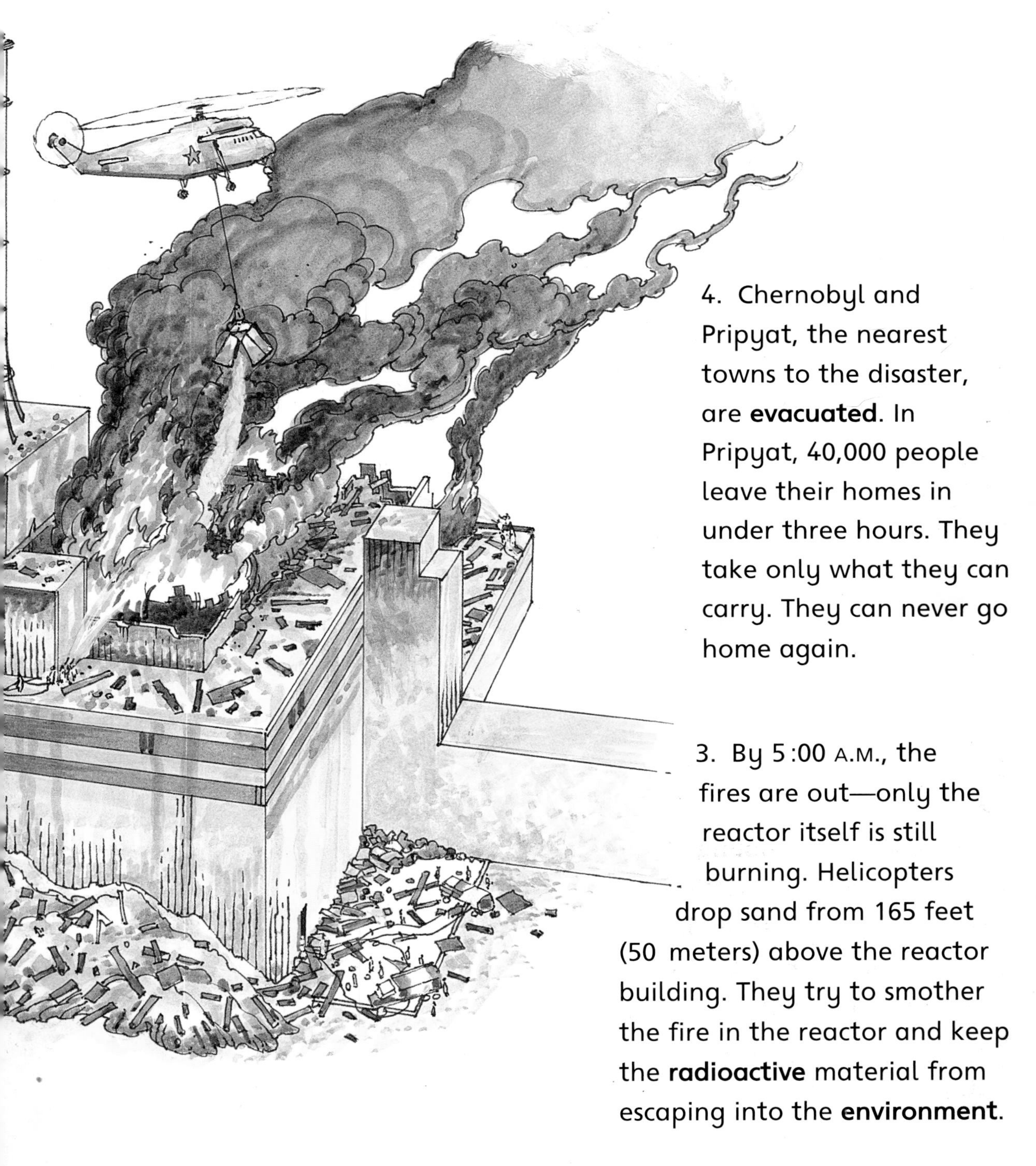

4. Chernobyl and Pripyat, the nearest towns to the disaster, are **evacuated**. In Pripyat, 40,000 people leave their homes in under three hours. They take only what they can carry. They can never go home again.

3. By 5:00 A.M., the fires are out—only the reactor itself is still burning. Helicopters drop sand from 165 feet (50 meters) above the reactor building. They try to smother the fire in the reactor and keep the **radioactive** material from escaping into the **environment**.

Deadly Secret

The fire in the Chernobyl **reactor** burned for ten days. It spewed 50 tons of deadly **radioactive** material into the air. Yet for two days the outside world knew nothing. The government of the **USSR** tried to keep the worst disaster in nuclear history a secret. It wanted no criticism of its **nuclear power stations** or of its safety policies.

The first warning to the outside world came on April 28. Workers at a Swedish nuclear power station measured frightening levels of **radiation**. The radiation was coming from the USSR. At 9 P.M., a four-line news item about Chernobyl was carried on Russian TV. The "accident" was official. But the secrecy had exposed hundreds of thousands of people to increased risks.

Who was to blame?

Six workers at the Chernobyl power station were put on trial for the disaster. The six workers were sentenced to a total of 40 years in prison for their careless neglect of safety. The Russian court put most of the blame on the managing director of the power station, Viktor Bryukhanov. At his trial, Bryukhanov said, "With so many deaths I can't say I'm completely innocent."

Thousands of people waited for buses to leave Chernobyl.

Workers built this giant concrete coffin over the destroyed reactor at Chernobyl nuclear power station.

Impossible clean-up

How do you clean up after a nuclear disaster? At Chernobyl, the first job was to keep more radioactive material from escaping. Once the fires were out, soldiers built a huge concrete coffin to enclose the reactor. Miners tunneled under the reactor to build a giant saucer. It kept the reactor from melting into the ground and **contaminating** the water of the whole region. Then over 600,000 people started to clean up the area. But a complete clean-up was impossible.

Dead land

A circle of land 38 miles (64 kilometers) around Chernobyl power station is still radioactive. It will remain closed to the outside world for 150 years. Only those who still work at the Chernobyl power station—where other reactors are still running—visit the area. Today, governments in Europe still believe Chernobyl is unsafe. They want it closed.

Lessons from Disaster

Ever since the first **nuclear power station** was built in the 1950s, people have feared accidents. Most accidents have not been serious, but some may have been kept secret. Governments and power companies do not like to admit that they have come close to disaster. They fear the reaction of the public.

The dangers of secrecy

Secrecy is one of nuclear power's big problems. In 1986, Chernobyl authorities kept the accident secret for several days, even though this put thousands of people at risk. At Windscale in 1957, four days passed before local people were warned not to drink milk **contaminated** by **radiation**. This history of secrecy has had some serious consequences—many people do not trust nuclear power stations or the people that run them. After the accident at Three Mile Island, people who had been **evacuated** from the site were suspicious of government claims that it was safe to go home. They simply didn't believe that any government would tell the truth about a nuclear accident.

Nuclear power stations, like this one in Sellafield, England, do not pollute the air with smoke. They can still be dangerous, though, if safety measures are not followed.

Is it safe?

Supporters of nuclear power say that new **reactor** designs are safer. Accidents are becoming less likely. Nuclear power, when it works properly, is less damaging to the **environment** than burning coal or oil to make electricity. And what else will we burn to make electricity when the world's supply of coal, oil, and gas runs out?

Anti-nuclear protesters warn of the dangers of nuclear power. They stress the fact that there are safer and cleaner ways of making electricity. There is also the problem of nuclear waste, which remains **radioactive** for thousands of years.

Many environmental organizations, such as Greenpeace, are opposed to nuclear power because of nuclear waste and the risk of accidents. This protest took place at Doodewaard nuclear power plant, the Netherlands, in 1994.

Future risks

Many older reactors, still in operation, are dangerous. Most are in the countries that made up the **USSR**. Some governments in Europe and the U.S. are paying to have the old reactors closed. In 1986, James Asseltine, then head of the U.S. Nuclear Regulatory Commission, said, "I have to advise **Congress** that there is a 45 percent chance of another serious nuclear accident within the next 20 years."

The World's Worst Nuclear Disaster

Chalk River, Canada, December 12, 1952 Partial **meltdown** in **reactor core**. No leaks, but the first known nuclear accident.

Kyshtym, USSR, 1957 (exact date kept secret) Nuclear waste container explodes, and 30 villages disappear from Russian maps. It is estimated that more than 8,000 people die as a result over the next 32 years.

Windscale (now called Sellafield), England, October 7, 1957 Fire in the reactor core causes a **radiation** leak. Estimated 33 deaths over the following years.

Idaho Falls, Idaho, January 3, 1961 Steam explosion in an experimental military **reactor** kills three soldiers.

Browns Ferry, Alabama, March 22, 1975 Fire in reactor wiring. No leak.

Three Mile Island, Pennsylvania, March 28, 1979 Partial meltdown in reactor core causes a radiation leak. No deaths, but the danger to local people is still unknown.

Tsuruga, Japan, March 8, 1981 Radioactive water leaks from reactor. Workers receive high doses of radiation.

Gore, Oklahoma, January 4, 1986 Worker killed by bursting container of nuclear material. Radiation leak leads to 100 people being admitted to hospital. No known deaths.

Chernobyl, USSR, April 26, 1986 Meltdown of reactor core. More than 200 people are killed, and perhaps 200,000 people more are at risk from radiation.

The map to the right shows the countries that have **nuclear power stations** and how many they have.

Radiation Sickness

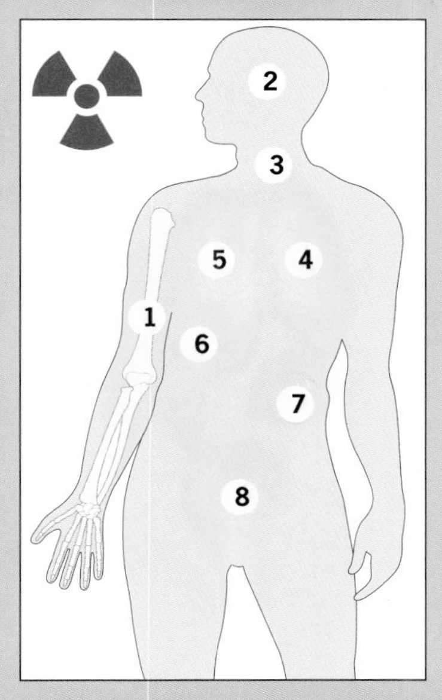

1. bone marrow damage (people lose the ability to fight disease)
2. unborn children suffer brain damage
3. thyroid cancer
4. lung cancer
5. breast cancer
6. liver failure
7. kidney failure
8. future children may suffer diseases

People exposed to high levels of radiation become sick. Many die within days or months. Others develop **cancer** and other diseases later in life.

On the nuclear map

There are over 400 nuclear reactors in the world. The United States, France, Japan, and Great Britain have the largest number of reactors.

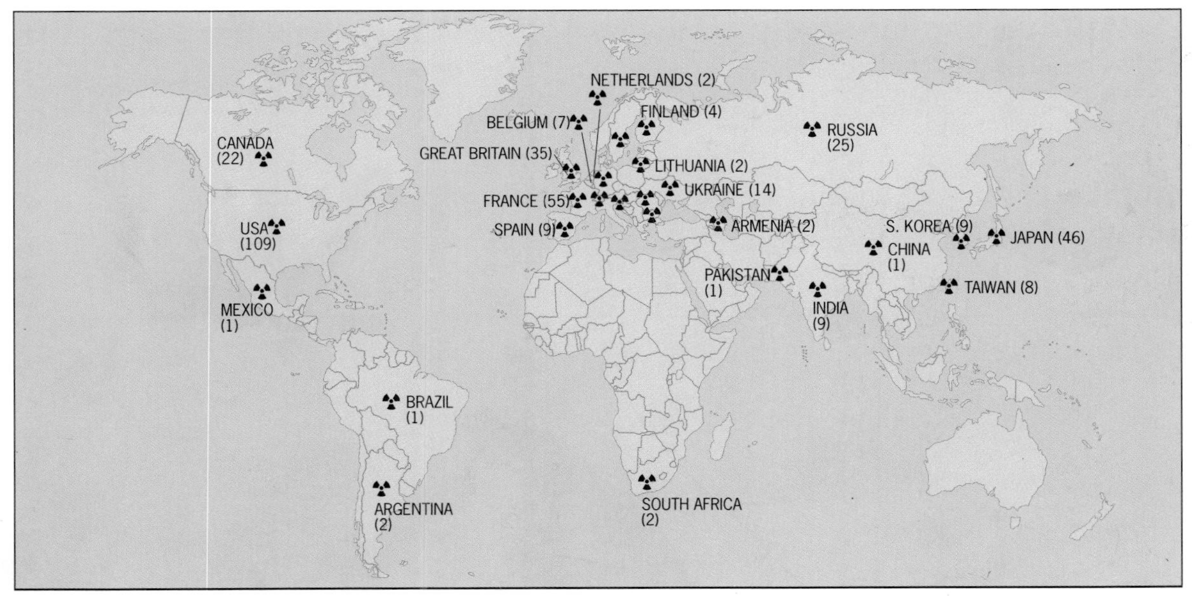

Glossary

cancer disease in which a group of cells grows and multiplies rapidly, killing nearby tissue. It often kills.

Congress branch of the U.S. government that makes laws

contaminated having a dangerous substance on or in it

control rods rods pushed in and out of a reactor to speed it up or slow it down

cooling system pipe and tanks that circulate water around a reactor or other machine to keep it cool

environment external surroundings; the land, water, and air around us

evacuate to move people away from a dangerous place until it is safe for them to return

fission splitting atoms by firing other atoms into them

fuel rods metal rods, often uranium, used as fuel inside a reactor

generator machine that makes electricity when it is turned

graphite rods rods, like very thick pencil leads, that control the power of the reactor (see control rods)

ignite set on fire

meltdown when a reactor burns and turns into a red hot liquid

molten melted and turned to hot, sticky liquid

national emergency serious emergency when emergency services nationwide should be on standby, ready to help if necessary

nuclear power station electricity generating station which uses nuclear energy

plutonium poisonous, radioactive metal element produced inside nuclear reactors, used in nuclear bombs

prototype working model of a new invention

radiation energy moving in the form of electromagnetic waves

radiation sickness illnesses people get when they receive large doses of radiation, including vomiting and burns

radioactive giving off particles and rays of radiation

reactor large tank or building holding the pipes and machinery needed to make nuclear energy

reactor core center of a reactor, holding fuel rods and control rods

turbine machine driven by a stream of gas, water, or steam

uranium gray metal element, heavier than lead, that is used as fuel in nuclear reactors

USSR (Union of Soviet Socialist Republics) communist country which included Russia and other nations. It separated in 1991.

valve small flap or door inside a pipe that turns on or off like a tap

More Books to Read

Condon, Judith. *Chernobyl & Other Nuclear Disasters.* Austin, Tex.: Raintree Steck-Vaughn Publishers, 1999.

Hamilton, Sue L. *Chernobyl: Nuclear Power Plant Explosion.* Minneapolis: ABDO Publishing Company, 1991.

Wilcox, Charlotte. *Powerhouse: Inside a Nuclear Power Plant.* Minneapolis: The Lerner Publishing Group, 1999.

Index